# CONSENSUS FOR
# THE LAY-UP OF BOILERS,
# TURBINES, TURBINE CONDENSERS,
# AND AUXILIARY EQUIPMENT

## AN ASME RESEARCH REPORT

**prepared by the**
TURBINE/TURBINE CONDENSER LAY-UP TASK GROUP
AND
BOILER LAY-UP TASK GROUP

**for the**
WATER TECHNOLOGY SUBCOMMITTEE OF THE
ASME RESEARCH AND TECHNOLOGY COMMITTEE
ON WATER AND STEAM IN THERMAL SYSTEMS

THE AMERICAN SOCIETY OF MECHANICAL ENGINEERS
Three Park Avenue ■ New York, New York 10016

# ◻ PREFACE ◻

The Water Technology Subcommittee of the ASME Research and Technology Committee on Water and Steam in Thermal Systems, under the leadership of Robert T. Holloway of Holloway Associates has established a "Consensus for the Lay-up of Boilers, Turbines, Turbine Condensers and Auxiliary Equipment." Since industry from time to time faces the need to idle equipment, the importance of good lay-up practices cannot be over-emphasized. Failure to properly protect equipment has frequently resulted in very high maintenance costs and, in some instances, the need to replace badly corroded equipment. These practices, properly applied, will aid the user in protecting these critical units from damaging corrosion during non-operating periods.

Data compilation and preparation of the resulting consensus on operating practices was performed by task groups of the subcommittee. The task groups consisted of a cross-section of manufacturers, operators, and consultants involved in the fabrication and operation of industrial and utility boilers, turbines and turbine condensers. The members of these task groups as well as those making significant contributions to this document are listed in the Acknowledgments section.

It is the intention of the ASME Research and Technology Committee to review this information, revise, and reissue it as necessary to comply with the advances in equipment design and lay-up technology. This document supersedes the "Consensus of Current Practices for Lay Up of Industrial and Utility Boilers" published in 1985 by the Industrial Subcommittee of the ASME Research Committee on Water in Thermal Power Systems.

It is my pleasure to commend all who generously contributed their time to this effort to improve the reliability of industrial steam generation equipment.

**David E. Simon II**
Past Chairman, ASME Research and Technology
Committee on Water and Steam in
Thermal Systems

# ◻ ACKNOWLEDGMENTS ◻

This document was prepared by the Turbine and Turbine Condenser Lay-up Task Group and the Boiler Lay-up Task Group for the Water Technology Subcommittee of the Research and Technology Committee on Water and Steam in Thermal Systems of the American Society of Mechanical Engineers. In addition, the former Utilities Subcommittee made significant contributions to the preparation of this document. Recognition is hereby given to the following members of these groups for their contributions in preparing this document.

**R. T. Holloway**, Chairman

## Combined Lay-up Document Task Group

| | |
|---|---|
| W. Allmon | H. Klein |
| A. Aschoff | R. W. Lane |
| A. Banweg | P. J. Latham |
| T. Beardwood | R. Long |
| J. Beecher | R. W. Light |
| J. Bellows | L. Paul |
| W. Bernahl | T. H. Pike |
| D. M. Bloom | J. O. Robinson |
| I. Cotton | L. S. Rosenzweig |
| B. Cunningham | J. J. Schuck |
| D. Daniels | K. Shields |
| P. Daniel | S. Shulder |
| D. Dewitt-Dick | J. Siegmund |
| S. B. Dilcer | D. E. Simon II |
| J. C. Dromgoole | J. Taylor |
| A. W. Fynsk | K. A. Teeters |
| F. Gabrielli | P. Thomasson |
| S. Goodstine | T. J. Tvedt, Jr. |
| R. Hackel | A. Whitehead |
| K. W. Herman | J. F. Wilkes |
| R. T. Holloway | W. Willsey |
| O. Jonas | D. K. Woodman |
| K. Kelley | |

# ❑ Safety Precautions ❑

The following precautions are intended to assist readers, who may be inexperienced with equipment lay-up, by outlining some hazards that may be encountered during lay-up procedures. They do not constitute a complete, detailed safety procedure. It is the responsibility of the personnel planning lay-up to be aware of the hazards, all OSHA, government and individual company regulations, etc., and to plan accordingly.

Proper handling of lay-up chemicals is essential from both a safety and an environmental standpoint. Some chemicals are flammable or explosive and can cause serious injury or death if mishandled. Other chemicals are toxic. Exposure may occur through skin contact or inhalation of vapors. All persons working with or near these chemicals must be familiar with their hazardous properties and recommended safety precautions.

Occupational Safety and Health Administration (OSHA) regulations, or individual company safety standards, should always be observed when handling nitrogen, which is an inert gas commonly used to blanket equipment during lay-up. It provides corrosion control when used to displace air from equipment. Nitrogen will not support human life and special precautions must be observed when it is used. Do not enter equipment laid up with nitrogen until the nitrogen supply is disconnected, the vessel is purged, and tests show that sufficient air is present and can be maintained.

Lay-up of equipment frequently requires confined space or vessel entry. Company procedures for testing and assuring the adequate oxygen content and absence of flammable gases must always be observed before any confined space is entered even if nitrogen gas is not used during lay-up.

Material Safety Data Sheets and any other available safety precautions must be provided for all chemicals that are used.

Wherever hazardous or unfamiliar chemicals are used, signs should be posted to warn of the dangers involved. Attention to warnings could very well save the lives of those entering a nitrogen-blanketed vessel. This is also the case when these chemicals are to be discharged. Warnings need to clearly indicate the expected dangers.

Before chemicals are released, it should be made certain that they have been deactivated or neutralized and present no danger to the environment. It may be necessary to notify the proper regulatory agencies prior to discharge.

Hydrazine, a common volatile oxygen scavenger, has been classified as toxic and a suspected carcinogen. Consult the chemical supplier's MSDS sheet for further precautions when using this chemical.

**Caution:** *When corrosion occurs in confined spaces, hydrogen is evolved creating a potentially explosive environment. Extreme caution should be exercised and such spaces must be checked and ventilated.*

# ❏ CONTENTS ❏

# ❑ SECTION 1 ❑
# INTRODUCTION

Steam and water equipment systems generally employ materials that readily corrode or pit when exposed to air and moisture and this is a condition often created by poor lay-up procedures. Thus without perceiving what is taking place, a user can unwittingly suffer loss of unit performance, impair serviceability, extend start-up, or cause an after start-up failure. Proper lay-up procedures can extend the useful life of a unit, reduce repair, replacement and maintenance costs and even prevent costly outages.

From a stand-by corrosion viewpoint, corrosion can attack the small single-stage steam turbine to the same degree as a large, multistage power generation turbine. Whether the plant is an industrial power facility or a power generation plant, the basic principle of keeping the combination of moisture and oxygen away from the metal surfaces is the same regardless of the treatment regime. This basic principle is easy to state, however, several different procedures may be employed to reduce or prevent stand-by corrosion. One method strives to eliminate the air while an alternate method strives to eliminate the moisture. For example, boilers can be filled with condensate to minimize oxygen exposure, while steam turbines have seals that allow leakage. Thus, steam turbines cannot hold liquid without modifications. Therefore, the preferred method for steam turbines is to minimize moisture. Steam condensers are normally laid up dry. Each unit also has its own special needs and requirements, thus the many lay-up guidelines available vary in approach, timing and application.

The intent of this document is to provide a consensus of current practices with resources and general guidelines, leaving development of the detailed procedure to the user. Only the user and associated technical support staff can access all the various factors such as availability, environmental constraints, atmospheric conditions, cost and other elements that need to be considered to arrive at the best solution with maximum benefits and protection, at minimum cost. While this document provides much useful lay-up information for boilers, steam turbines and related equipment, qualified personnel should still be involved in the selection of the final lay-up procedure used.

This document addresses most components found in a steam generating system for electrical power generation as well as mechanical drive services. Regardless of the lay-up method used, the basic philosophy is the same — to keep the combination of oxygen and moisture away from parts that can corrode. As mentioned above, there can be a basic differ- ence in how this may be accomplished in the static components of the system (boiler) and the rotating components of the system (steam turbine). The boiler components can be laid up using either a wet or dry procedure while the rotating machinery such as the steam turbine and the steam condenser must be laid up in a dry condition.

The practices described herein have been proven in field use. The practices described for the boiler are in agreement with Section VII of the ASME Boiler and Pressure Vessel Code. They are applicable both to new boilers before being placed in service and to operating boilers during an outage.

Types of boilers covered are:
    (1) Industrial-fossil fired
    (2) Industrial-waste heat
    (3) Marine
    (4) Utility-fossil fired*
    (5) Gas turbine heat recovery steam generators (HRSG)

---

*This document describes procedures, etc., commonly used for drum type boilers. Procedures used for once through boilers are similar but the equipment manufacturer should be consulted.

# □ Section 2 □
# FACTORS TO BE CONSIDERED

The following is a list of items which generally need to be addressed when preparing to lay-up a unit or system;

(1) Safety — of personnel and property
(2) Systems and/or equipment to lay-up
(3) Condition of unit(s)
(4) Estimated period of lay-up and work to be performed
(5) The lay-up environment such as possibility of freezing
(6) Estimated time available to start up unit
(7) Availability of manpower to inspect and monitor unit(s) during lay-up
(8) Value of unit and possible financial consequences of lay-up procedure selected
(9) Compatibility between lay-up and operational chemical treatments
(10) Disposal of lay-up solutions
(11 System auxiliaries such as fuel additive feed systems, soot blowers etc.

# ⊓ Section 3 ⊓
# REASONS FOR LAY-UP

During any non-operating periods after initial hydrotest, boilers and related equipment must be laid up properly to avoid corrosion damage. Failure to do so will reduce their availability, reliability, increase maintenance costs, and eventually shorten their useful life. Severe corrosion could even lead to catastrophic failure.

After any welding is done to piping on the boiler it is often hydrostatically tested or "hydro" tested. It should be noted that the water used for hydro testing of boilers and related equipment should be of the same purity and treated with the same chemicals used for lay-up.

Steam turbines are particularly susceptible to corrosion damage because air is usually drawn into the turbine during cool down. The oxygen in the air combines with steam that has condensed in low spots and cavities that are not drainable, thus providing the ingredients for corrosion to quickly develop. The presence of chlorides and sulfates in the condensate adds to corrosivity. If deposits are present on metal surfaces, they can promote the formation of under-deposit corrosion cells that may result in localized corrosion (pitting).

In boilers, both the water/steam sides and the fire/gas sides must be protected. If the waterside of a boiler is left open to the atmosphere unprotected, corrosion will occur particularly at liquid-air interfaces. Corrosion can also occur in the pre-boiler and after-boiler sections.

In addition to the local damage that occurs, the corrosion products generated in the pre-boiler section will migrate to the boiler, steam turbine, reheater, condenser and condensate system when the system is returned to service. These, along with the corrosion products generated in the boiler, may deposit on critical heat transfer areas of the boiler, increasing the potential for localized corrosion and/or overheating during operation. If they don't deposit in the boiler, they will probably continue downstream to damage the superheaters and steam turbines, particularly the blading where high velocities and corrosion products can cause severe erosion damage in a short time.

## ⊓ 3.1 Steam Turbine Steam Side and Boiler Water Side Factors

The items listed below are factors that contribute to corrosion on the steam side of steam turbines and water side of idle boilers and related equipment.

(1) *Moisture*--Under completely dry conditions, corrosion is negligible.

(2) *Dissolved Oxygen*--Corrosion rates are related to the dissolved oxygen concentration in a moist environment. Any conditions that increase the oxygen concentration in moisture, or allow the continued addition of oxygen, will aggravate the corrosion process.

(3) *Low pH*--Low pH of water generally increases the tendency of metals to corrode.

(4) *Deposits*--Presence of deposits on steam or waterside surfaces may cause the development of corrosion cells that promote localized corrosion during lay-up periods.

(5) *Contaminants*--In general, the greater the contaminant levels (such as chloride and sulfates) in water containing oxygen, the faster the corrosion rate.

## 3.2 Boiler Fire Side Factors

The conditions listed below are known to cause, or at least contribute to, corrosion on the fire and gas sides of
non-operating steam generators:

(1) Presence of absorbent deposits;

(2) Moisture absorbed in such deposits;

(3) Corrodents such as sulfurous and sulfuric acids which may form and attack steel under deposits;

(4) Oxygen dissolved in moisture in deposits

# □ Section 4 □
# IDENTIFICATION OF
# MATERIALS IN EQUIPMENT

Identifying the materials in any system to be laid up is key to developing a suitable lay-up procedure. The reader should be aware that some materials used in the various components of steam plant equipment are not compatible with some chemicals that may be considered for the lay-up procedure. For example, high concentrations of ammonia can cause problems with copper alloys. The following procedures must be modified accordingly.

Generally, the materials utilized should be detailed in the manufacturer's equipment manuals but there are many variations and combinations of materials available. It is the reader's responsibility to determine the metallurgy utilized in the system, identify components that might be affected, and to consult qualified personnel to clearly establish suitability and/or compatibility.

# □ SECTION 5 □
# INSPECTION OF EQUIPMENT

The *ASME Boiler and Pressure Vessel Code, Section VII, Recommended Guidelines for the Care of Power Boilers* has a Subsection on Inspection (C6) that details the requirements for the boiler proper. Paragraph C6.200 notes that "Most jurisdictions require that all power boilers should be given one or more internal and external inspections per year by an Authorized Inspector as defined in C6.100(a)." Another Subsection on Control of Internal Chemical Conditions (C8) has a paragraph on Inspection (C8.570) that is quoted here in part as it adds useful information to the subject of lay-up.

"When boiler outages permit, thorough inspection should be made of all the internal parts of the boiler to note the presence of corrosion in any form and to permit the collection of deposits for later examination for corrosion products. Where any case of unusual and extensive corrosion is noted, the services of qualified personnel may be secured to determine causes and to suggest remedial changes in water treatment and operation." This suggestion may be applied as well to the inspection of equipment ahead of the boiler such as feedwater heaters, deaerators and deaerator storage tanks.

As noted above, when equipment is out of service there is an opportunity to inspect, make repairs and perform routine maintenance. Equipment may also be taken out of service for economic reasons. In either case, proper lay-up is critical. Damage can occur quickly to systems that are not properly treated. Regardless of the circumstances, recording what was inspected, what was done and what is to be monitored can identify useful low-cost steps that can be taken to circumvent later repairs. These steps could pay large dividends in money and time if or when the system is returned to service.

Consequently, the recommended procedure is for a complete inspection of the unit before lay-up when the waterside is breached. Then decisions on cleaning, repair and/or replacement can then be accurately formulated. As noted in the ASME Code, "a record of each inspection should be kept in a uniform manner so that any change of condition can be definitely noted and compared, especially with reference to the thickness of scale, corrosion, erosion, cracks, and other unusual conditions." (ref. Sect. VII, Subsection C6.630). Using the services of a manufacturer's representative should also be considered to provide technical assistance on equipment inspections and formulation of cleaning solutions, repairs and lay-up procedures for major system components and/or support systems.

The post-inspection report should be detailed, including photographs and deposit analyses, and should specify the condition of the equipment, comparing present condition with that during previous inspections. This procedure will help determine if deposits are new or unusual. The report may also define the type of analyses to be performed, during future inspections, such as soluble iron and pH, which are especially important for turbine deposits, but which are not routinely requested.

For the steam turbine, inspection is not easy, as it usually requires removal of the cover or top half in order to inspect the internals such as blading, diaphragms, seals and rotor. Specific disassembly procedures and inspection criteria are normally found in the owner's manual or unit's technical manual.

For the turbine condenser, both the steam condensate and the cooling water sides should be inspected. While all of these items also apply to the steam turbine, additional items such as internal clearances, condition of bearings, and balance of rotor should also be recorded. Properly identified photos of machine deposits, blade erosion and bearing condition can also be very useful. If deposits are unusual, chemical analysis of deposits should be performed.

# ☐ SECTION 6 ☐
# CLEANING OF EQUIPMENT

Boilers must be properly removed from service to minimize adherence of suspended solids to boiler metal surfaces. Suspended solid adherence can be reduced by immediate flushing with hot, pressurized water, while waterside surfaces are still moist. A boiler containing deposits that formed during operation or because of improper removal from service may require chemical cleaning at some time during the outage period.

Any chemical cleaning procedures should be designed and supervised by trained and qualified personnel to ensure successful deposit removal and to avoid potential equipment damage. It is important that a protective metal oxide film be reestablished on the metal surfaces immediately following chemical cleaning. This film can be formed most effectively after chemical cleaning if the boiler is operated for a short period of time following the cleaning. One recommended procedure is to operate with a solution of passivating chemicals after chemical cleaning to obtain metal passivation before lay-up. If this cannot be done, it is recommended that the chemical cleaning be scheduled immediately preceding restart. It is recommended that boiler fireside deposits be removed prior to lay-up due to the potential for under-deposit corrosion. Samples of deposits should be collected for analysis.

A dirty turbine should not be placed in lay-up. Cleaning of the steam turbine is usually done with the turbine open and most often by mechanical means (blasting). The equipment manufacturer normally recommends a schedule for inspection and cleaning of a steam turbine. Normally, the only cleaning required for the steam side of the steam turbine condenser is for flushing off of all accumulated matter with condensate or other high purity water. On the other hand, the cooling water side may require extensive cleaning. Flushing may be adequate, but mechanical or chemical cleaning is often required.

It is important that equipment be placed in a fully protective lay-up after cleaning as described in the remainder of this document.

# ☐ Section 7 ☐
# USE AND DISPOSAL OF
# LAY-UP CHEMICALS

The following factors should be considered in using chemicals for lay-up:

(1) All normal safety precautions associated with using the chemical(s) should be observed.

(2) If the lay-up chemical(s) is different from the operating chemical(s), either in type or concentration, the lay-up solution may have to be replaced or the concentration reduced before start-up.

(3) Disposal of lay-up chemical solutions must comply with all applicable state and federal regulations. Environmental considerations may limit the use of concentrated solutions, or solutions with low levels of volatile oxygen scavenger, for wet lay-up.

# ❏ Section 8 ❏
# BOILER LAY-UP — WATERSIDE

## ❏ 8.1 Short-term versus Long-term Lay-up of Equipment

The information on "Reasons for Lay-up" provided in Section 3 of this document, is also applicable to this section.

Previous documents have defined the duration of lay-up as "short" or "long" outages sometimes defined by a specified number of days applicable to distinct classes of equipment. This document does not define the duration closely but allows the user to select an applicable procedure based on plant circumstances, conditions, operating requirements, and complexity of the system.

Boilers in "hot standby" or "banked" are not covered by this document as such conditions are considered to be operational. If such a condition changes to lay-up as the result of operating demand, the proper lay-up procedures should be determined and implemented. Even if the boiler is in a "hot standby" condition, lay-up of steam-side of the condensers, turbines, and other equipment may be appropriate. These lay-up procedures may be found in Sections 10-13 of this document.

"Short" term lay-up is intended to cover those circumstances where equipment must be taken off line for inspection, maintenance, repair etc., where it is intended to place the equipment back into service as rapidly as possible. Generally this period would not exceed one or two weeks.

"Long" term lay-up is intended to cover circumstances where it is clear the equipment must be out of service for an extended period of time for maintenance, seasonal need, etc. A unit in a long-term lay-up will typically have at least a few days notice prior to restarting.

Lay-up procedures are specified separately for simple and complex piping systems. Simple systems are those where there is no superheater or where the superheater can be drained or flushed. Complex systems are those that have a non-drainable superheater and/or reheater.

This document does not cover the condition, commonly referred to as "mothballing," where equipment is permanently removed from service but maintained in a preserved state for possible return to service if ever required.

Regardless of which piece of equipment is laid up or for how long, the primary need and concern is for isolation. This aspect cannot be overemphasized for a variety of reasons, SAFETY being foremost. Other reasons include contaminant prevention, the avoidance of moisture addition, and the overall success of lay-up in preserving the equipment. Inadequate isolation during an outage has probably contributed more to equipment deterioration than any other factor.

The ASME Code requires two isolation block or stop valves, with an ample free-blow drain valve between them, for all boiler outlets when connecting to a system. The two stop valves should be closed and the drain opened whenever the boiler is taken out of service. This effectively ensures that the boiler is isolated from the rest of the system.

While not stated in the Code, the most probable purpose of the drain is to bleed off pressure and ensure that the boiler is not passing flow into the system or the system is not passing steam flow into the boiler. However, this arrangement can be very effective at providing isolation in that any leakage past the valve under pressure flows out the drain and identifies a leaking valve, provided one can see the drain outlet. Regardless of the piece of equipment to be laid up, this principle can and should be used.

For equipment such as steam turbines, deaerators and steam jet air ejectors that are to be laid up, it is critical for the laid-up equipment to be isolated from any and all active steam systems. While blinds can be used very effectively, they are time consuming to install and remove. They also need to be clearly marked and documented for future restart planning. From an equipment lay-up viewpoint, a small steam leak into a unit can negate all the preparatory work and inflict substantial damage on a piece of equipment in a very short time. In systems with condensing turbines, the condenser and steam turbine are usually considered as one block for lay-up purposes.

While the code is explicit where multiple boilers feed into a common system, it does not address systems outside the boiler proper and therefore makes no comments on valve arrangements for associated steam equipment.

A good arrangement for unit isolation is a "double block and bleed system" as previously described for the boiler outlet. This might consist of gate valve plus globe valve main line isolators with a free-blow drain valve between the two block valves. When the two block valves are closed and the drain valve is open, a visual check can reveal if there is any leak across the supply side isolating valve.

This may be contrasted with "single block and bleed" arrangement sometimes used for the downstream, or exhaust side, of a turbine. The drain valve is trapped to remove condensate from the piping system downstream of the isolating gate valve. However, this arrangement does not allow for visual observation that the exhaust side isolator is leaking.

Any leakage across the isolator would allow steam to flow back into the turbine. Considerable corrosion damage might result when no means of leak detection is provided. Use of a second isolator or "block" valve between the steam turbine exhaust relief valve and the turbine exhaust gate valve isolator, combined with a free-blow drain valve between the two isolators, would provide positive isolation and a means of checking the success of isolation.

Paragraph 2.2.6 of ANSI / ASME publication, B19. 3c-1989, " Safety Standard for Compressors for Process Industries" states important points regarding unit isolation and safety during lay-up. The following is paraphrased from that document and should be implemented only after the turbine is cooled on turning gear.

When maintenance or lay-up is being performed on rotating equipment, compressors, turbines, generators etc., precautions shall be taken to ensure isolation of all energy sources from the driver or system. These precautions shall include the use of blinds or double block valves and bleeder on steam and fuel gas supplies to drivers, or in the case of motors and generators, either:

(1) Electrical load centers shall have a switching arrangement that must be locked in the open position, tagged and tried; or

(2) Other positive means of current interruption shall be employed.

(3) In all cases, all connected equipment shall be depressurized to prevent rotation of the drive shaft.

## ☐ 8.2 Boiler Lay-up — Waterside

The two generally accepted lay-up methods for the waterside of the boiler are the dry and the wet procedures. In wet procedures, the boiler is left full of water with either standard boiler water or an alkaline, reducing solution. A pressurized nitrogen blanket is maintained, or suitable alternatives used, to prevent oxygen ingress. In dry procedures, the boiler is completely dried and maintained in the dry condition during the period of lay-up.

Often when a boiler is taken down, it is to repair something in the boiler such as a leaking tube. These cases typically require that the system be drained to allow access to the repair area. Following the repairs, the boiler and superheater are hydro-tested. The water used for the hydro-test should be feedwater quality, preferably deaerated, and chemically treated. Once the hydro-test is completed, the boiler should be laid up immediately with one of the wet lay-up procedures described in this document.

If outage periods are to be short and/or frequent, and the boiler is known to be relatively free of sludge and other internal deposits, draining and inspection may be omitted. In this case, vigorous, manual boiler blowdown should be utilized and increased several days prior to shutdown to maximize sludge removal.

The recommended lay-up procedures (dry and wet) are summarized in Sections 8.2.1 and 8.2.2. In Section 8.3.4 a Decision Tree can be found to assist the reader in determining the correct lay-up method. This is followed by detailed recommended lay-up procedures in Section 8. 4 and 8.5.

The procedures are specified separately for simple and complex piping systems. Simple systems are those where there is no superheater or where the superheater can be drained or flushed. Complex systems are those that have a non-drainable superheater.

### 8.2.1 ■ Dry Lay-Up

Dry lay-up is most commonly used for simple circuitry boilers during a long outage. The objective of dry lay-up is to keep metal surfaces free of moisture containing dissolved oxygen. The waterside of industrial, marine and waste heat boilers containing simple circuits may be laid up dry. Dry lay-up is not recommended for boilers with complex circuitry because it is difficult to get and keep all internal surfaces free of moisture.

There are several methods used for dry lay-up. In one dry procedure, the boiler is vented, drained and dried as completely as possible with air. Then manholes are closed, all connections are tightly closed or blanked, and the boiler is purged and pressurized with nitrogen to prevent ingression of air. The success of this procedure depends on the rapid and thorough drying of all boiler metal surfaces and the exclusion of air during lay-up. An alternative and perhaps better method is to drain under nitrogen pressure to prevent air ingress during draining. In this case, it is not necessary to dry the boiler completely prior to establishing the nitrogen blanket. It is important to remember that when using a nitrogen blanket, the nitrogen pressure must be maintained at all times for protection to be complete.

Another alternative for maintaining an air-dried boiler in a dry condition utilizes desiccants installed in suitable locations to maintain a moisture-free atmosphere on the water side. Related equipment such as economizers, deaerating heaters, and feedwater heaters may be laid up in a similar manner.

A boiler containing porous moisture-retaining deposits should not be laid up dry because of the difficulty in completely drying out the deposits and the potential for subsequent under-deposit corrosion. The recommended dry lay-up procedures are discussed beginning in Section 8.5.

## 8.2.2 ■ Wet Lay-Up

*Chemicals Used*

Chemicals used for wet lay-up are generally classified into two categories: non-volatile inorganic and volatile. Both types of lay-up chemicals are effective when applied properly. However, lay-up chemicals used should typically be consistent with the operating chemical regime. Non-volatile inorganic chemicals are commonly used in low and medium pressure boilers that either do not have superheaters or have drainable superheaters (simple circuits) and operate on softened water make-up. Volatile chemicals are commonly used in medium and high-pressure boilers with complex piping circuits that operate on demineralized water make-up. Oxygen scavengers and pH control chemicals are the commonly used materials and are shown below with their respective chemical concentrations:

### Table 1 - Commonly Used Lay-up Chemicals

|  | Non-volatile Inorganic Chemicals | Volatile Chemicals |
|---|---|---|
| Oxygen Scavengers-Note 2 | sodium sulfite | hydrazine - Note 1 |
| pH control chemicals | sodium hydroxide | ammonia or a neutralizing amine |

### Table 2 - Boiler Water Chemical Concentrations During Lay-up

|  | Non-volatile Inorganic Chemicals | Volatile Chemicals |
|---|---|---|
| Oxygen Scavengers-Note 2, 4 | 200 mg/L (ppm) as $Na_2SO_3$ | 100-200 mg/L (ppm) min. hydrazine as $N_2H_4$-Note 1 |
| pH control chemicals | 400 mg/L (ppm) min. hydroxide alkalinity as $CaCO_3$ | Note 3 |

**Notes to Tables 1 & 2:**

(1) If an oxygen scavenger other than hydrazine is used, the user should confirm the suitability and concentration of the substitute chemical with a qualified water treatment consultant.

(2) Either catalyzed or uncatalyzed oxygen scavengers may be used.

(3) Sufficient chemical is required to maintain a minimum of 10.0 pH for all-ferrous systems or a pH of 9.3-9.5 for systems containing copper alloys. To achieve a pH of 10 in high purity water, for example, requires approximately 10.0 mg/L (ppm) of ammonia (as $NH_3$) or 12. 4 mg/L (ppm) of cyclohexylamine (as $C_6H_{11}NH_2$). Other organic amines will require different concentrations to achieve the desired pH.

(4) Oxygen scavengers need not be used for lay-up if they are not used in the normal feedwater treatment program.

Nitrogen for lay-up should contain a minimum of 99.6% nitrogen. This purity is typically found in commercial-grade liquid nitrogen and compressed nitrogen gas cylinders.

Recommended lay-up maintenance and testing practices can be found in Section 14 of this document.

*Preparation of Lay-up Solution*

Deaerated water of feedwater quality should be used for laying up boilers. Raw surface water or other waters containing significant suspended matter should not be used. For boilers with non-drainable superheaters and reheaters, utility boilers, and other high-pressure industrial boilers that use high quality feedwater, only high quality condensate or deaerated demineralized water and volatile chemicals should be used to prepare lay-up solutions.

*Preparation and Filling with Lay-up Chemical--All Boilers*

The basic objective of the following procedures (1–4) is to achieve a uniform concentration of the lay-up chemicals throughout the equipment.

Prior to filling any boiler equipment the water temperature and metal temperature of headers and waterwalls must typically be within 100-150°F of the water being added to prevent thermal shock to the system. Consult the OEM for specific recommendations regarding these temperatures. The following paragraphs, 1–4, outline several possible procedures that are suitable for the preparation and addition of lay-up solutions to economizers, boilers, superheaters and reheaters.

(1) Prepare the chemical solution in a separate tank. Using a suitable metering pump, blend the lay-up chemicals at the concentrations required in the boiler into the fill water as it enters the economizer and boiler.

(2) If condensate quality is acceptable and contamination is not a cause for shutdown, an alternative method would be introduction of concentrated lay-up solutions directly into the condensate receiver, with mixing to achieve the desired concentration.

(3) Prepare the chemical solution in a separate tank and add it to the boiler while filling. The solution can be added to the steam drum through an existing chemical feed line.

(4) After the boiler is filled and the lay-up solution has been added, fire the boiler lightly for at least 30 minutes to obtain circulation and distribution of chemicals. The economizer must then be filled with properly treated water when the pressure has decreased to about 5 psig (34 kPag). The high water alarm should be deactivated and all vent valves operated as needed so that the boiler can be filled completely. After filling, the boiler and economizer must be isolated by tightly closing the feedwater valve. If firing is not possible, the rate of addition of both the lay-up solution and the fill water must be very carefully controlled to ensure uniform distribution of treated water throughout the boiler.

### Boilers With Superheaters and/or Reheaters

Special precautions are required for boilers with non-drainable superheaters and reheaters.

(1) All chemicals used for lay-up of non-drainable superheaters and reheaters must be volatile to prevent deposit formation in these components when the boiler is returned to service. Superheaters should be back-filled with chemically treated demineralized quality water to minimize deposition or corrosion after firing. Preferably, volatile chemicals should also be used for laying up the economizer and boiler. If non-volatile inorganic chemicals are used in the boiler, it is essential that precautions be taken so that none of the boiler lay-up solution enters the superheater.

(2) For some steam-cooled surfaces (superheaters, reheaters), it may be necessary to fix/pin supports (as for hydro) before filling them with water.

(3) If non-drainable superheaters and reheaters are constructed of austenitic stainless steel, it is recommended that the demineralized water used for filling have a conductivity of $\leq 1$ $\mu S/cm$.

### Excluding Air

In all wet lay-up procedures it is important to exclude air to eliminate any air/water interface where the presence of oxygen could cause increased corrosion in a concentrated area.

To prevent inleakage of air (oxygen) during the lay-up period, pressurize with 5 psig (34 kPag) nitrogen through a suitable connection.

Where the boiler is completely filled with water instead of using a nitrogen blanket, it may be necessary to install an expansion vessel (filled with properly treated water) above the highest part of the boiler to accommodate volume changes caused by temperature changes in the boiler. The vessel should be fitted with a cover and a sight glass and connected to an available opening such as a vent line at the top of the boiler to create a hydrostatic head. The tank will provide a ready visual check of water level loss or inleakage during lay-up. This tank must be disconnected prior to start-up (see Section 15).

For boilers with complex circuitry (non-drainable superheater): After the boiler, superheater, and reheater are filled with lay-up solution, all valves on connecting piping should be tightly closed and the equipment pressurized with 5 psig (34 kPag) nitrogen.

## ☐ 8.3 Procedure Selection

### 8.3.1 ■ Reasons for Lay-Up

The reasons for lay-up are detailed in Section 3 of this document and it is recommended that the section be read again. The first paragraph is repeated here for convenience.

During any non-operating periods after the initial hydrotest, boilers and related equipment must be laid up properly to avoid corrosion damage. Failure to do so will reduce their availability, reliability, increase maintenance costs, and eventually shorten their useful life. Severe corrosion could even lead to catastrophic failure.

### 8.3.2 ■ Using Decision Tree

The reader should use the following definitions and the Glossary of Terms, Section 17 at the end of this document, to define the lay-up parameters. First determine whether the duration is short or long. Then determine whether the boiler system will be breached or will not be breached during lay-up. Finally, determine if the boiler circuitry is simple or complex. The reader should use the definitions of these terms in the Glossary of this document.

### 8.3.3 ■ Definition of Terms in Decision Tree

Boilers in "hot standby" or "banked" are not covered by this document as such conditions are considered to be operational. If such a condition changes to lay-up as the result of operating demand, the proper lay-up procedures should be determined and implemented.

This document does not cover the condition, commonly referred to as "mothballing," where equipment is permanently removed from service but maintained in a preserved state for possible return to service if ever required.

"Short" term lay-up is intended, in this section, to cover those circumstances where equipment must be taken off line for inspection, maintenance, repair, lack of demand, etc., where it is intended to place the equipment back into service as rapidly as possible. Generally this period would not exceed one or two weeks.

"Long" term lay-up is intended, in this section, to cover circumstances where it is clear the equipment must be out of service for an extended period of time for maintenance, repair etc. A unit in a long-term lay-up will typically have at least a few days notice prior to restarting.

### 8.3.4 ■ Decision Tree

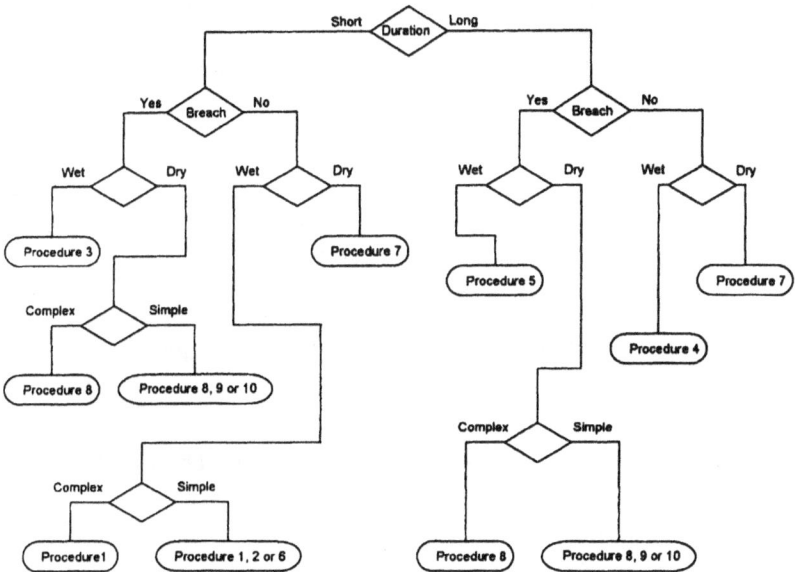

**Figure 1 Lay-Up Decision Tree**

## ❐ 8.4 Waterside Wet Lay-up Procedures

### 8.4.1 ■ *Procedure 1— Short Outage — No Breach of Waterside System — Simple and Complex Circuits*

**Nitrogen Cap**

This procedure is also applicable to systems following a hydro-test. If freezing is a problem apply auxiliary heat. Alternatively, for simple circuits use a dry lay-up procedure. **Caution:** *The boiler must not be in a confined area such as in an enclosed building as vents or valve packing leaks could displace oxygen in the space.*

(1) As a minimum, maintain the same oxygen scavenger and pH control chemical concentrations as those present during normal operation. Any additional water added to the boiler as it cools must be deaerated and chemically treated.

(2) Establish a 5 psig (34 kPag) nitrogen cap on the superheater and the steam drum through the drum vent and superheater outlet header drain/vent as the unit is cooled and before a positive pressure is lost. Boiler vents should not be opened to atmosphere. Avoid the admission of air into any water-containing circuit during lay-up. Once the nitrogen blanket is established, it is critical that it is constantly maintained or corrosion can result.

(3) Complete lay-up of the steam cycle requires lay up of auxiliary equipment such as feedwater heaters, condensers and deaearators. Lay-up procedures for this equipment can be found in Sections 10 and 12.

### 8.4.2 ■ *Procedure 2 — Short Outage — No Breach of Waterside System — Simple Circuits Only*

**Water Cap**

This procedure is also applicable to systems following a hydro-test. If freezing is a problem apply auxiliary heat. Alternatively, for simple circuits use a dry lay-up procedure.

(1) Maintain the same oxygen scavenger and pH control chemical concentrations as those present during normal operation.

(2) Fill boiler completely with feedwater up to the vent. Drainable superheaters should be backfilled with deaerated feedwater containing volatile oxygen scavenger. To accommodate volume changes caused by temperature changes, connect an expansion vessel filled

with properly treated water to the drum vent. The vessel should be located above the highest part of the boiler and connected such that a hydrostatic head can be maintained on the boiler at all times. The vessel should be fitted with a cover and a sight glass. The tank should be monitored regularly and will provide a ready visual check of water level loss or inleakage during lay-up. This tank should be drained to waste prior to startup.

(3) Complete lay-up of the steam cycle requires lay-up of auxiliary equipment such as feedwater heaters, condensers, and deaearators. Lay-up procedures for this equipment can be found in Sections 10 and 12.

### 8.4.3 ■ Procedure 3 — Short Outage — Breach of Waterside System — Simple and Complex Circuits

**Nitrogen Cap**

Since the system will be breached, dry lay-up is the appropriate lay-up procedure for all affected equipment, until maintenance is complete. When complete, the boiler can be left in a dry lay-up condition or filled preparatory to start-up. If filled, the following wet lay-up procedure applies. If freezing is a problem apply auxiliary heat. Alternatively use a dry lay-up procedure. **Caution:** *The boiler must not be in a confined area such as in an enclosed building as vents or valve packing leaks could displace oxygen in the room.*

(1) Drain and open only those sections requiring repairs.

(2) Isolate remainder of unit under 5 psig (34 kPag) nitrogen pressure where it can be done safely. Avoid the admission of air to isolated areas.

(3) Maintain the same oxygen scavenger and pH control chemical concentrations for water remaining in the cycle as those present during normal operation.

(4) Once maintenance and the hydrostatic test are complete, the boiler should be filled to the normal operating level with treated and deaerated boiler feedwater and a 5 psig (34 kPag) nitrogen blanket established on the steam drum. If deaerated water cannot be used, higher concentrations of oxygen scavenger and amine (or other pH control chemical) must be added to the boiler fill water as specified in Procedure 4.

(5) Complete lay-up of the steam cycle requires lay-up of auxiliary equipment such as feedwater heaters, condensers, and deaearators. Lay-up procedures for this equipment can be found in Section 12.

### 8.4.4 ■ *Procedure 4 — Long Outage — No Breach of Waterside System — Complex Circuits*

**Chemical Protection with Nitrogen Cap**

If freezing is a problem apply auxiliary heat to areas of the boiler prone to freezing. **Caution:** *The boiler must not be in a confined area such as in an enclosed building as vents or valve packing leaks could displace oxygen in the room.*

(1) Add sufficient volatile oxygen scavenger and volatile neutralizing amine or ammonia to the waterwalls, economizer, and associated piping to protect it against oxygen. The recommended level for hydrazine is 200 mg/L (ppm) for an all-ferrous system and 100 mg/L (ppm) for feedwater piping that contains copper. The amine or ammonia concentration should be such that the pH of the water is 10 for an all-ferrous system and 9.3-9.5 for mixed metallurgy systems. If other oxygen scavengers and amines are used, consult the chemical vendor for the proper dosage.

(2) Volatile oxygen scavenger and volatile neutralizing amine or ammonia should be added in a manner that results in a uniform concentration throughout. They may be added to the system in several ways, for example:

(a) Pump concentrated solutions through the chemical feed equipment and blend-fill to achieve the desired concentrations.

(b) If condenser leakage is not a cause for shutdown, concentrated solutions can be introduced directly into the hotwell where they can be mixed to achieve the desired concentration and pumped forward. If ion exchange condensate polishers are employed, they must be bypassed during this operation.

(3) It is important to have the water temperature in the boiler below 204°C (400°F) before addition of volatile oxygen scavenger. If this temperature is exceeded, some volatile oxygen scavengers, such as hydrazine, will start to decompose. Also, temperature differential between the metal temperatures and the added water temperature cannot exceed the manufacturer's recommendations.

(4) The tube side of copper alloy feedwater heaters should be filled with condensate-quality water containing volatile neutralizing amine or ammonia to adjust the pH to 9.5 and 50 to 100 mg/L (ppm) of volatile oxygen scavenger.

(5) The superheater can be simply blanketed with nitrogen empty or back-filled with condensate quality water containing 200 mg/L (ppm) of volatile oxygen scavenger or performance equivalent and 10 mg/L (ppm) of volatile neutralizing amine or ammonia. Add the

fill water to the outlet of the non-drainable section allowing the water to fill the superheater completely as indicated by a rise in drum level.

**CAUTION:** *Steam-cooled surfaces (superheaters, reheaters) will typically require additional structural support prior to being filled with water.*

(6) Establish and maintain a 5 psig (34 kPag) nitrogen cap on the superheater and steam drum. Nitrogen cap the shell side of the feedwater heaters. Nitrogen blanket should be applied through the drum vent and superheater outlet header drain/vent as the unit is cooled when pressure drops to 5 psig (34 kPag). Avoid admission of air to all water-containing circuits throughout the duration of the lay-up.

(7) Test lay-up solution weekly for the first month and then monthly thereafter. If the volatile oxygen scavenger concentration decreases by more than 33% of the original level or if the pH of the boiler water drops to below 9.0, additional chemicals must be added to the boiler to re-establish the desired concentrations. Refer to section 14 for additional instructions on monitoring.

(8) Complete lay-up of the steam cycle requires lay-up of auxiliary equipment such as feedwater heaters, condensers, and deaerators. Lay-up procedures for this equipment can be found in Section 12.

### 8.4.5 ■ *Procedure 5 — Long Outage — Breach of Waterside System — Complex or Simple Circuits*

**Chemical Protection with Nitrogen Cap**

Since the system will be breached, dry lay-up is the appropriate lay-up procedure for all affected equipment, until maintenance is complete. When complete, the boiler can be left in a dry lay-up condition or filled preparatory to start-up. If filled, the following wet lay-up procedure applies. If freezing is a problem apply auxiliary heat. **Caution:** *The boiler must not be in a confined area such as in an enclosed building as vents or valve packing leaks could displace oxygen in the room.*

(1) Drain and open only those sections requiring repairs.

(2) Fill the remainder of the boiler with the recommended concentrations of the appropriate pH control and oxygen scavenging chemicals consistent with the normal feedwater chemical treatment.

   (a) For complex circuits this is typically a volatile oxygen scavenger such as hydrazine (or equivalent) and volatile neutralizing amine or ammonia such that the concentration in the waterwalls, economizer and feedwater heaters (tube side) and associated piping are 200 mg/L (ppm) and 10 mg/L (ppm), respectively for all ferrous systems and 100 mg/L (ppm) and pH of 9.3-9.5 for mixed metallurgy systems.

(b) For simple circuits, where non-volatile chemicals are used for operation, the lay-up chemicals should be sodium hydroxide and sodium sulfite such that the concentrations in the boiler are 400 mg/L (ppm) hydroxide alkalinity as $CaCO_3$, and 200 mg/L (ppm) as $Na_2SO_3$, respectively.

(3) Lay-up chemicals should be added in a manner that results in a uniform concentration throughout. They may be added to the system in several ways, for example:

(a) Pump concentrated solutions through the chemical feed equipment and blend-fill to achieve the desired concentrations.

(b) If condenser leakage is not a cause for shutdown, concentrated solutions can be introduced directly into the hot well where they can be mixed to achieve the desired concentration and pumped forward. If ion exchange condensate polishers are employed, they must be bypassed during this operation.

(4) It is important to have the fluid temperature in the cycle below 204°C (400°F) before addition of volatile oxygen scavenger. If this temperature is exceeded, some scavengers, such as hydrazine, will start to decompose. Also, temperature differential between the metal temperatures and the added water temperature cannot exceed the manufacturer's recommendations.

(5) The superheater can be blanketed with nitrogen or back-filled with condensate-quality water containing 200 mg/L (ppm) of volatile oxygen scavenger or performance equivalent and 10 mg/L (ppm) of volatile neutralizing amine or ammonia. The pH of the solution should be approximately 10.0. Add the fill water to the outlet of the non-drainable section allowing the water to fill the superheater completely as indicated by a rise in drum level. A nitrogen cap of 5 psig (34 kPag) should be established on the steam drum of the boiler to prevent air inleakage.

**CAUTION**: *Any steam-cooled surface (superheater, reheater) will typically require additional structural support prior to being filled with water.*

(6) Establish and maintain a 5 psig (34 kPag) nitrogen cap on the superheater and steam drum. Nitrogen cap the shell side of the feedwater heaters. Nitrogen blanket should be applied through the drum vent and superheater outlet header drain/vent as the unit is cooled when pressure drops to 5 psig (34 kPag). Avoid admission of

air to all water-containing circuits throughout the duration of the lay-up.

(7) Once the chemical concentrations are established, regular testing must be performed to ensure that treatment levels are maintained. Test lay-up solution weekly for the first month and then monthly thereafter. If the volatile oxygen scavenger concentration decreases by more than 33% of the original value, or if the pH of the boiler water drops to below 9.0, additional chemicals must be added to the boiler to re-establish the desired concentrations. Recommendations on testing can be found in Section 14.

(8) If chemical additions are made, it is important to ensure sufficient mixing so that the concentration of the scavenger and amine are uniform. This can be achieved by firing the boiler lightly or by an external circulation pump. Attention should be directed to valve maintenance that may cause treated water loss or the addition of untreated water into the boiler and auxiliaries causing dilution of properly treated lay-up solutions.

### 8.4.6 ■ Procedure 6 — Short Outage — No Breach of Waterside System — Simple Circuits Only

**Steam Cap**

This procedure is not applicable to boilers with superheaters. Air (oxygen) ingress can be prevented by keeping the waterside of the boiler under slight constant pressure. The pressure must remain above atmospheric at all times for this procedure to be successful.

(1) The continuous blowdown water from a properly treated operating boiler can be piped into a mud drum header on the idle boiler, allowing the water to overflow from the steam drum vents to an acceptable disposal location.

(2) Steam-heated coils can be installed in the mud drum to maintain temperature in the boiler or the boiler may be intermittently fired to maintain a positive pressure. An oxygen-free, treated feedwater source is necessary to replenish the water lost through steam production.

### ❒ 8.5 Waterside Dry Lay-up Procedures

There are a number of methods for creating and maintaining a dry lay-up condition in a boiler. These procedures discuss the use of a continuous nitrogen purge, dehumidified air, and chemical desiccants as dry lay-up methods.

An alternative for small industrial boilers is use of silica gel as a static dehumidifier in a sealed boiler. The procedure for the Use of Desiccants (p.31) can be used to determine the amount of silica to be used and provides a recommended inspection schedule. If metal corrosion is observed, correct problem of moisture ingress. If desiccant is exhausted, dry out boiler again, replace with fresh silica gel.

### 8.5.1 ■ Procedure 7 — Long or Short Outage — No Breach of Waterside System — Simple and Complex Circuits

**Nitrogen Drain and Blanket**

**Caution:** *The boiler must not be in a confined area such as in an enclosed building as vents or valve packing leaks could displace oxygen in the room.*

(1) Generally remove boiler from service, drain the boiler under a nitrogen overpressure before the boiler pressure dissipates completely (e.g., at 10 psig). Liquid water must be drained from all sections of the steam/water circuit. Check all low point drains to ensure removal of as much moisture as possible.

(2) Maintain a nitrogen overpressure of 5 psig (34 kPag) throughout draining and subsequent storage. All circuits must be pressurized with nitrogen to exclude air (oxygen) ingress. The safety precautions discussed at the beginning of this document must be given careful attention.

### 8.5.2 ■ Procedure 8 — Long or Short Outage — Breach of Waterside System — Simple and Complex Circuits

**Nitrogen Drain and Blanket**

**Caution:** *The boiler must not be in a confined area such as in an enclosed building as vents or valve packing leaks could displace oxygen in the room.*

(1) Drain and open only those sections requiring repairs. Some sections may not be able to be drained completely nor blanketed with nitrogen during maintenance activities and some corrosion may occur. Therefore it is important that repairs be made quickly and that the nitrogen blanket be re-established as soon as possible.

(2) Where it can be done safely, isolate remainder of unit under 5 psig (34 kPag) nitrogen pressure. Avoid the admission of air (oxygen) to isolated areas.

(3) Once maintenance and the hydrostatic test are complete, drain all water-filled equipment under a nitrogen overpressure of 5 psig (34

kPag). Liquid water must be drained from all sections of the steam/water circuit. Check all low point drains to ensure removal of as much moisture as possible.

(4) Maintain a nitrogen overpressure of 5 psig (34 kPag) throughout draining and subsequent storage. All circuits must be pressurized with nitrogen to exclude air (oxygen) ingress. The safety precautions discussed at the beginning of this document must be given careful attention.

(5) Complete lay-up of the steam cycle requires lay-up of auxiliary equipment such as feedwater heaters, condensers, and deaerators. Lay-up procedures for this equipment can be found in Section 12.

### 8.5.3 ■ Procedure 9 — Long Outage — Breach of Waterside System — Simple Circuits Only

#### Recirculating Dry Air

*Note*: Waterwall surfaces must be clean to use this lay-up procedure.

(1) To facilitate drying of surfaces, it is preferable to drain the boiler hot, while the pressure is still ≤ 10 psig (68 kPag).

(2) While the boiler is still hot and before condensation can occur, (typically within 8 hours), begin circulating filtered dehumidified air through the boiler.

(3) Piping and dehumidification systems must be designed for the specific boiler layout. The system must be instrumented and inspected to assure reliable operation. This should include monitoring of the relative humidity of air leaving the boiler. The direction of flow and the bleed points are not important so long as the flow is through the entire system. The airflow should be sufficient to dry the system out in less than one week, i.e., the faster, the better.

(4) For the equipment to be considered dry, the air leaving the equipment should have a maximum dew point of -10°F (-23°C). After this dew point is reached on all the bleeds, the flow can be reduced to approximately 1% of the system volume per hour.

(5) Inspect the dehumidification equipment and boiler at regular intervals.

#### Use of Desiccants

This procedure is an alternative for small industrial package-type boilers with no superheater.

(1) To facilitate drying of surfaces, it is preferable to drain the boiler hot, while the pressure is still ≤ 10 psig (68 kPag).

(2) When boiler is cool, and after observing all safety practices for confined-space entry, install bags of a desiccant, in trays, at all avail-

able openings. Either silica gel or quicklime can be used, however the preferred desiccant is silica gel. The amount of desiccant used depends on the boiler size and is shown below

Silica gel —( 5.0 lbs/100 ft$^3$ (2.7 kg/m$^3$) of boiler volume. This is available with a color indicator to show the degree of moisture absorption.

Quicklime — 2.0 lbs/100 ft$^3$ (1.1 kg/m$^3$) of boiler volume.
(3) Close all manholes and blank or close all connections on the boiler as completely as possible to prevent ingress of moist air.
(4) Check the desiccant daily for the first week, then weekly for the next 7 weeks. Replace as necessary. All work should be performed as quickly as possible to minimize entry of moisture.
(5) Inspect the waterside of the boiler every three months for evidence of active corrosion.

**Note**: Spilled desiccant may cause corrosion where it comes in contact with metal surfaces. See Section 16 (10) for corrective actions to be taken.

# □ Section 9 □
# BOILER LAY-UP — FIRESIDE

When boilers are to be laid up, the fireside must be considered as well as the waterside. Two common methods used for fireside lay-up are identified as cold and hot. The preferred method for each installation will depend on local conditions.

The primary objective in fireside lay-up is to keep the metal surfaces dry. If appreciable amounts of deposits are present, the fireside should be cleaned before lay-up. This is especially true where fuels containing high concentrations of sulfur have been used.

## □ 9.1 Hot Lay-up

The following items should be considered where hot lay-up is planned.
(1) Since certain boilers cannot be effectively cleaned on the gas side without extensive dismantling, a hot lay-up may be necessary.
(2) The hot lay-up method may not be practical for extended periods because of energy costs.
(3) To prevent moisture absorption by deposits on the gas side, the temperature of the metal surfaces must be kept at 170°F (77° C) or higher, depending upon the sulfur content of the fuel.
(4) Fireside metal temperatures can be maintained at a safe level by controlled light firing with low sulfur fuel or by the use of electric hot air blowers.
(5) The circulation of hot blowdown water through the boiler waterside may not be sufficient to assure metal temperatures of 170°F (77° C) or higher in all parts of the boiler because of the parallel circuits involved and cooling resulting from air infiltration. However, the installation of a steam coil in the lower drum should provide sufficient heat input and circulation to produce adequate metal temperatures throughout the boiler.

## □ 9.2 Cold Lay-up

The following items should be considered where a cold lay-up is planned.
(1) If fuel containing high concentrations of sulfur has been used, fire the boiler, if practical, on natural gas or number 2 oil for one week prior to the shutdown. This technique may help remove

corrosive deposits, but its effectiveness should be verified by subsequent inspection of the fireside surfaces.

(2) Before removing the boiler from service, operate all soot blowers starting with the blower farthest from the stack.

(3) When the unit is off line and the water temperature has dropped sufficiently, wash down the gas side of the boiler, economizer, air heater, and the flue gas side of the ID fan. Use a five percent solution of an alkaline chemical such as soda ash to neutralize acidic deposits. Be sure that all wash water is completely drained from the equipment.

(4) After this procedure, all of the areas contacted by wash water should be dried by some technique such as firing with sulfur-free fuel.

(5) Moisture should not be allowed to remain in the refractory or insulation. The relative humidity of the air should be kept as low as practicable. Seal the furnace as completely as possible to minimize entry of moist air. Keep the metal surface dry by the use of heat lamps, dehumidification with desiccants along with air circulation, circulation of dry warm air, or a combination of these. Thicker deposits may require removal by mechanical cleaning.

(6) Make provisions to keep rainwater from entering the boiler through the stack.

(7) Inspect the fireside once per month for evidence of active corrosion and, if found, take corrective measures as previously discussed.

# ❑ Section 10 ❑
# STEAM TURBINE, TURBINE CONDENSER, AND REHEATER LAY-UP

This procedure is not applicable to air-cooled condensers. Consult the condenser manufacturer for specific procedures for lay-up of this equipment.

## ❑ 10.1 Steam Side

The steam turbine, turbine condenser and reheater are considered as a unit because they are normally interconnected with no reasonable means of isolating them from each other. If the reheater can be isolated from the high and intermediate pressure stages of the turbine, the steam turbine/condenser and the reheater can be laid up separately. Unlike the boiler, the steam side of the steam turbine can normally only be laid up dry. The steam turbine shaft rests in bearings, which are oil-lubricated and separated from the steam side by seals. These seals come in various arrangements but are generally not positive isolators nor designed for liquid, therefore the steam turbine should be laid up dry.

All potential sources of moisture intrusion such as from the extraction, backpressure or auxiliary sealing steam connections, must be eliminated by blanking or double-blocking and bleeding. Any moisture that enters the system will likely remain in the system and combine with other impurities to attack metal surfaces.

The steam turbine should be shut down as recommended by the manufacturer. Depending on the system and the manufacturer's recommendations, it is desirable to start lay-up procedures prior to complete cooling·of the unit and turning off the turning gear (if unit is so equipped). As soon as practical after removing steam supply from the unit, the steam turbine/condenser should be drained and dried as completely as possible.

Carry out the drying process by flowing either filtered, dehumidified air or nitrogen through the system. The direction of flow and the bleed points are not important so long as flow is through the entire system. If air is used, the quantity should be sufficient to dry the system out in less than one week, i.e., the faster, the better. If nitrogen is used the time can be considerably longer, as the oxygen content in the system should be negligible after a short time. For the equipment to be considered dry, the gas leaving the equipment should have a maximum dew point of –10°F (–23°C). After this dew point is reached on all the bleeds, the airflow can be reduced to approximately 1% of the system volume per hour.

**CAUTION**: *Nitrogen will not support human life and special precautions must be observed when it is used. Do not enter equipment laid up with nitrogen until the nitrogen supply is disconnected, the vessel is purged, and tests show that sufficient air is present and maintained. Ensure that nitrogen is not vented into an enclosed room or area.*

If the system is going to be down for more than two weeks, the lubrication system should be operated and the unit rotor turned, using the turning gear, or manually rolled. When rolling the shaft, the starting point should be marked to ensure that the rotor comes to rest at a new point approximately 90° from the previous resting point. This procedure should be repeated every two weeks.

If the steam turbine has not been inspected for some time and the turbine is going to be laid up more than two (2) months, the turbine and turbine condenser should be inspected and cleaned before lay-up procedures are implemented. It is also suggested the condenser shell side be inspected at the same time for steam impingement, corrosion at tube supports and the integrity of the baffles and the drain lines entering the unit. If the turbine or condenser is suspected of being dirty, inspection and cleaning, if necessary, is recommended regardless of the outage duration.

After the system has been opened and cleaned, long-term lay-up procedures can make use of either filtered, dehumidified, dry air or pressurized dry nitrogen. If filtered dry air is used, then a continuous flow must be maintained through the system. If dry nitrogen is used and it is possible to pressurize the system, a small positive pressure must be maintained. For this to be successful, the turbine rotor seals will probably require some modifications. Nitrogen for turbine lay-up should contain a minimum of 99.6% nitrogen. This purity is typically found in liquid nitrogen and compressed nitrogen gas cylinders.

If the rotating equipment is going to be laid up for an extended period of time and monitoring manpower is limited and the rotors are flexible shafts, consideration should be given to removing the rotors from the equipment. This would eliminate the need for periodic rotation of the rotor and/or complete sealing of the system. For long-term flexible shaft rotor storage many maintenance engineers prefer vertical hanging in a controlled atmosphere (low humidity). While this method of lay-up reduces the need for frequent monitoring of the equipment, it does increase both the time and labor to recommission the system and its cost.

## ❏ 10.2 Waterside of Turbine Condenser

### 10.2.1 ■ Short-Term Lay-Up
Two procedures are commonly practiced:
  (1) Cooling water is continuously circulated through the condenser tubes. Care must be taken to ensure that the cooling water contains an adequate concentration of corrosion inhibitors, biocides, and, if needed, dispersants. In most cases, the normal cooling water treatment program for microbiological and corrosion control can be continued; however the chemical vendor should be consulted. It is important to avoid stagnation in any areas of the cooling water system.
  (2) The condenser tube side is filled with cooling tower water containing the appropriate concentrations of treatment chemicals, and then isolated. If the recirculating cooling water has very high conductivity, it may be desirable to use water with lower Total Dissolved Solids and Specific Conductance for filling the Condenser. Higher than normal concentrations of non-oxidizing biocide and corrosion inhibitors might be considered.  Care must be taken to ensure that adequate protection is provided from corrosive and microbiological attack. A possible disadvantage of this method is that suspended solids can settle on metal surfaces and promote the formation of under-deposit corrosion cells. Problems associated with high suspended solids can be minimized by increasing the cooling tower blowdown rate prior to filling the condenser, thus purging much of the suspended solids from the cooling system. Procedure (2) is not suitable for systems using once through cooling water unless the source has low Total Dissolved Solids.

### 10.2.2 ■ Long-Term Lay-Up
For long-term lay-up, the water side of the turbine condenser should be inspected and cleaned, if necessary. It should then be dried by using filtered, dehumidified air. Drying is considered complete when the air leaving the condenser has a maximum dew point of $-10°F$ ($-23°C$).  When drying is complete, either the condenser should be pressurized with nitrogen to prevent air inleakage or the circulation of dehumidified air should be continued.

# ❒ SECTION 11 ❒
# LAY-UP OF TURBINE OIL SYSTEM

The turbine oil, including that in the reservoir, should be cleaned prior to lay-up. The cleanup must remove water and suspended matter since these impurities can support microbiological activity. The cleaned oil should be treated with a corrosion inhibitor and a biocide recommended by the oil manufacturer, then circulated for about a day to obtain good mixing of these materials with the oil and to establish a good protective film on the metal surfaces. During the lay-up period, the oil should then be circulated for several hours and tested on a scheduled basis, at least monthly. Tests should include moisture content, microorganism counts, contaminant levels, inhibitor level, particle count, and wear metals. If problems are found, corrective action must be taken.

# ❏ Section 12 ❏
# LAY-UP OF FEEDWATER HEATERS AND DEAERATORS

## ❏ 12.1 Lay-up of Feedwater Heaters

The tube side of feedwater heaters should be treated the same as the boiler during lay-up periods. The shell side normally is steam blanketed or dried and pressurized with nitrogen during lay-up. However, the shell side may be flooded with treated condensate or condensate-quality water. All-steel systems should be laid up with condensate containing 200 mg/L (ppm) volatile oxygen scavenger or equivalent, and ammonia or suitable amine such as cyclohexylamine as needed to maintain the pH at 10.0 min. For copper or copper alloy systems, use 50 to 100 mg/L (ppm) of volatile oxygen scavenger and adjust the pH to 9.5.

## ❏ 12.2 Lay-up of Deaerator and DA Storage Tank

A number of methods may be used to properly lay up the deaerator and deaerator storage tank. It is preferred to maintain a steam blanket on the equipment as this ensures some hot deaerated water for the subsequent startup. If steam is not available then the deaerator and storage tank may be:

(1) Blanketed with a small continuous flow (30 SCFH) of nitrogen
(2) Drained while hot and maintained dry with dehumidified air or desiccant
(3) Filled to the vent with water containing volatile oxygen scavenger and either ammonia or amine, as previously described for feedwater heaters

# ☐ Section 13 ☐
# LAY-UP OF AUXILIARY COMPONENTS AND PRE-BOILER SYSTEMS

## ☐ 13.1 Scope

The protection of the auxiliary components of a power plant during stand-by is equally as important as the protection of the boilers and/or turbines. However, it may be impractical to protect all auxiliary equipment during a shutdown. For example, it may require more time to protect some auxiliary parts than the length of time of the shutdown. In other cases, the cost of treatment may approach the cost of replacement, in which case stand-by treatment would not be cost effective. Plant personnel must decide which components are critical.

Depending upon the nature of the equipment and the anticipated length of shutdown, the equipment can be stored wet or dry. In either case, the objective is to prevent damage of the auxiliary equipment due to corrosion and/or by freezing.

Whenever a specific piece of equipment is removed from service, where available, the manufacturer's instructions should be followed.

Electrical equipment should be isolated and locked out as required for safety purposes. Where equipment is outdoors, precautions should be taken to prevent freezing.

## ☐ 13.2 Inspection of Equipment

It is desirable that all critical equipment be inspected before being placed into lay-up. The inspection should include the integral parts of the equipment including both sides of heat exchange equipment. Whenever possible, detailed photographs of the equipment should be taken and a record made of the condition of the equipment. When the outage is short-term, inspection may be impractical.

## ☐ 13.3 Lay-up Term Definition for Auxiliary Equipment

For the purpose of protecting the auxiliary equipment, short-term lay-up is very generally defined to mean that it is expected that the equipment will be restarted soon (no longer than two weeks) and it is desired to maintain a state of readiness while minimizing any possible damage. For some equipment, "short-term" may represent a different time period than for other equipment.

Long-term lay-up implies no state of readiness and the main concern is for prevention of corrosion damage.

As stated in Section 8.1, this document does not cover the condition, commonly referred to as "mothballing," where equipment is permanently removed from service but maintained in a preserved state for possible return to service if ever required. Many auxiliary components, such as listed below, are not normally subjects of either "short" or "long" term lay-up procedures except under special circumstances such as "mothballing."

- Motors and generators
- Pumps
- Fans
- Induced draft fans
- Piping
- Air receivers, inert gas tanks and other similar tanks
- Tanks
- Ejectors, chillers, etc.
- Switch gear, circuit breakers, control cabinets, relay cabinets, etc.
- Conveyors, hoppers and other handling equipment

## ❐ 13.4 Chemical Feeding Equipment

The equipment manufacturer's instructions should be followed where available. Every chemical feed system is different. The following suggested procedure may not cover every situation and should be regarded only as a guide.

### 13.4.1 ■ Solution Feeders — Excluding Demineralizer Regeneration Systems

(1) Flush chemical day tanks, pumps, piping, valves, injection lines and instruments associated with the chemical feed systems with condensate quality water. Drain and allow to dry.

(2) Chemicals may remain in bulk tanks. However, plant personnel should review each chemical for storage suitability.

(3) Dry chemicals may remain in storage in accordance with the storage suitability.

(4) Solution (Dissolving) tanks should be drained and the contents stored in marked drums or disposed of in accordance with government guidelines. The tanks should be flushed with condensate quality water and the tanks dried and sealed for a very long lay-up.

### 13.4.2 ■ Dry Feeders

(1) Close the isolating valves of slide gates between the dry chemical hoppers and run the dry chemical out of the feeder.

(2) The chemical day tanks, pumps, piping, valves, instruments, etc. associated with the chemical feed systems should be flushed with clean water and drained before lay-up.

(3) Dry chemicals may remain in the bulk tank during short-term lay-up in accordance with the storage instructions for each chemical.

(4) Solution (Dissolving) tanks should be drained and the contents stored in marked drums or disposed of in accordance with government guidelines. The tanks should be flushed with condensate quality water and the tanks dried and sealed for a very long lay-up.

### 13.4.3 ■ Pretreatment Equipment

Frequently, there is not enough manpower available to carry out a detailed short-term lay-up procedure for pre-treatment equipment. As a result, most plants do little or nothing with regard to short-term shutdown. However, there are some things that must be considered with each shutdown.

Wherever available, the manufacturer's instructions should be consulted and followed. In lieu of specific manufacturer's instructions, the following suggestions may apply. However, keep in mind that few pre-treatment systems are the same type or configuration depending on influent water quality and the required effluent purity. The following suggestions cannot cover every situation and should be regarded only as a guide.

*Ion Exchange Equipment*

(1) Ion exchange resin should be kept wet, protected from freezing, and should not be subjected to temperatures greater than 95°F (35°C). After ion exchange resins have been in service, they may also be subject to biological growths and should be protected from biological action by a micro-defouling cleaning procedure prior to storage. The resin or equipment vendor can supply a cleaning procedure.

(2) If lay-up is expected to be greater than two weeks, all ion exchange resins should be regenerated normally prior to shutdown.

(3) Although the stability of strong acid cation resins in the regenerated hydrogen form is greater than that of strong base anion resins in the regenerated hydroxide form, for long-term chemical stability reasons, consideration should be given to conversion of strong acid cation resins to the sodium form for storage. Similarly, strong base anion resins should be converted to a more stable form such as the chloride, sulfate or carbonate form. Detailed and recommended conversion procedures should be obtained from the resin manufacturers.

(4) The following is a rough guideline applicable to sodium or chloride conversion only. Exhaust the resin using a 5% to 10% by weight sodium chloride solution. The quantity of sodium chloride solution required for this is four times the total volume of the ion exchange resin.

(5) If the planned lay-up will extend beyond twelve months, consideration should be given to converting the resin to the sodium and chloride form as in 3 above, unloading and wet-storing the complete resin charge. When returning resins to service, stored ion exchange resin requires double or triple regenerations before normal effluent quality can be achieved. Depending on the resin volume and value, it might be more cost effective to consider disposal of the resin following the appropriate governmental regulations and the purchase of new resin when the equipment is to be placed back into service. It may be advantageous to purchase resin in the regenerated form so that the unit can be put back into service quickly.

(6) Ion exchange vessels may be drained, dried, and stored in place if indoors and not subjected to extreme temperatures. Ion exchange vessels that must be stored outdoors and/or in direct sun should be covered with tarps and stored full of water. In this case, a vent should be left open to prevent pressure from building up from heat and the water level in the vessel should be checked and re-established periodically. Care must be taken to prevent freezing as this can destroy vessel internals.

(a) Empty lined ion exchange vessels should not be stored in direct sun nor in areas in which the temperatures can become greater than 95°F (35°C).

(b) Ion exchange vessels with stainless steel internals that have been used to prepare ion exchange resin for storage as outlined above, should be rinsed with water with a low chloride content before either wet or dry storage.

(c) Lined ion exchange vessels should not be returned to service after an extended lay-up without the integrity of the lining being confirmed by spark testing.

(7) Acid and caustic storage tanks associated with the demineralizer will require proper lay-up and storage. Consult the bulk chemical supplier for proper procedures.

### Reverse Osmosis (RO) Equipment

"Short-term" lay- up, with RO membranes in place in pressure tubes, is defined as more than 5 days and less than 30 days out of operation. "Long-term" lay-up, with RO membranes in place in pressure tubes, is defined as more than 30 days out of operation.

Before undertaking any "short" or "long" term lay-up operation, the membrane manufacturer should be contacted for specific instructions related to the specific equipment and local environments.

For thin film composite (Polyamide) membrane elements
    (1) Short-term lay-up
        (a) Flush the RO system with feedwater, while simultaneously venting any gas from the system.
        (b) When the pressure tubes are filled, close the appropriate valves to prevent air from entering the system.
        (c) Reflush as described at 5-day intervals.
    (2) Long-term lay- up
        (a) Clean the RO membrane elements in place.
        (b) Flush the RO section with a solution of permeate that contains an appropriate, approved biocide.
        (c) When the RO section is completely filled with this solution, close the valves to retain the solution in the RO section.
        (d) Repeat steps b and c with fresh solution every 30 days if the temperature is below 80°F (27°C) or every 5 days if the temperature is above 80°F (27°C).
        (e) When the RO system is ready to be returned to service, flush the system for approximately 1 hour using low-pressure feedwater with the product dump valve open to drain; then flush to drain at high pressure for 5 to 10 minutes. Before returning to service, ensure there is no residual biocide in the product water.

For cellulose acetate blend membrane elements
    (1) Short-term lay-up
        (a) Clean the RO membrane elements in place.
        (b) Flush the RO section with acidified water (pH 5.0–6.0) that contains a free chlorine residual of 0.1–0.5 mg/L (ppm) when measured at the brine outlet.
            (i) When the RO section is completely filled with this solution, close the valves to retain the solution in the RO section.
            (ii) Repeat steps b and c every 2 days.
            (iii) The membrane manufacturer may provide identification of other approved biocides that do not require such frequent flushing.
    (2) Long-term lay-up
        (a) Clean the RO membrane elements in place.
        (b) Flush the RO section with a solution of permeate that contains an appropriate, approved biocide with the pH adjusted to 5.0-6.0.
        (c) When the RO section is completely filled with this solution, close the valves to retain the solution in the RO section.

(i) Repeat steps b and c using fresh solution every 30 days if the temperature is below 80°F (27°C) or every 5 days if the temperature is above 80°F (27°C).

(ii) When the RO system is ready to be returned to service, flush the system for approximately 1 hour using low-pressure feedwater with the product dump valve open to drain; then flush to drain at high pressure for 5 to 10 minutes. Before returning to service, ensure there is no residual biocide in the product.

## Dry storage of membranes

(1) RO elements should only be stored dry as supplied prior to installation. They should be protected from direct sunlight and stored in a cool, dry environment with an ambient temperature range of 68°–95°F (20°–35°C).

(2) Precautions should be taken to avoid damage caused by freezing water in mechanical equipment, pumps, piping, valves, instruments, etc., associated with the reverse osmosis system that have been flushed as a result of preparing the membranes as above. For short-term lay-up this equipment may remain with water in place.

(3) RO systems generally have chemical feed equipment and membrane cleaning systems. These should be flushed with clean water and drained before extended lay-up.

## Granular Media Filters

(1) The filter media should be backwashed for an extended period at high flow rate to ensure cleanliness before shutting the filter system down.

(2) The water need not be drained from the filter vessels unless there is a danger of freezing. However, if the shut-down will be 3 months or more, an appropriate biocide should be used to prevent microbiological growth in the bed. Alternatively, the media should be removed for disposal.

## Deep Bed Condensate Polishers

See "Ion Exchange Equipment" (p. 46).

## Powdered Resin Condensate Polishing Equipment

(1) For short periods of time, powdered resin may be stored in the polisher vessels. Each powdered resin condensate polishing vessel generally has a holding pump to recycle water from the vessel outlet to the vessel inlet in order to hold the powdered resin on the filter elements. For short-term lay-up, this pump may be used to maintain the polishing system in a ready condition.

(2) In general, powdered resin for condensate polishing systems is prepared just before it is required. The powdered resin system should

be stored in accordance with the instructions of the manufacturer. Lacking those instructions, the following procedures may be considered:

(a) Flush all powdered ion exchange resin from the polisher vessels.
(b) Empty polisher and slurry vessels should be rinsed with water with a low chloride content before storage.
(c) All vessels and tanks should be drained and stored dry and in place if indoors and not subjected to extreme high temperatures and protected from freezing.

## Cold Lime Softeners and Clarifiers

For short- and long-term lay-up, the water may remain in the cold lime softener, clarifier and thickener tanks.

(1) Chemical feed should be terminated.
(2) The sludge removal system, rake, blowdown valve, thickeners and/or sludge dewatering equipment should continue to operate until all sludge has been processed out of the system. For very short duration, sludge in the unit can be retained by maintaining sludge fluidity by continuous operation of the recirculating and sludge rake system.
(3) Water flow to the unit should be shut off.

## Hot Process Softeners

For short- and long-term lay-up, water may remain in the hot process softener tank as long as steam supply can be maintained.

(1) Chemical feed should be terminated.
(2) The sludge removal system, blowdown valves, thickeners and/or sludge dewatering equipment should continue operating until all sludge has been processed out of the system. For very short durations, up to 2 days, sludge may be retained in the unit by continuous circulation, at high flow rate, to the unit inlet.
(3) Water flow to the hot process softener should be shut off.
(4) After the sludge is processed out of the system, all electrical equipment should be turned off. The equipment may be left in place during a short-term lay-up. If the equipment is outdoors, it should be protected from the weather.

# ❏ Section 14 ❏
# LAY-UP MONITORING
# AND MAINTENANCE

Maintaining lay-up integrity is essential to a successful program that maintains the asset and provides for return to service of a reliable unit when needed without unbudgeted expense. The result of inadequate maintenance can be serious corrosion damage to equipment. The lay-up program should be carefully monitored to ensure that corrosion control is maintained. For example, the loss of pressure within a closed vessel indicates a leak. If pressure is lost, moist air can enter the equipment and form condensate on metal surfaces.

Condensate can absorb oxygen and carbon dioxide at the air / water interface and corrode the metal surfaces. If atmospheric chloride also enters the equipment, the combination of moisture, oxygen, carbon dioxide, and chloride can result in an extremely severe corrosive attack to the metal. Therefore, all leaks should be found and promptly fixed. When filtered dehumidified air is used for lay-up, the supply flow should be monitored as well as the relative humidity and dew point of the air leaving the equipment. Deviations from the set and normal levels need to be addressed and corrected.

## ❏ 14.1 Testing of Lay-up Solutions

Lay-up solutions should be tested weekly until the rate of loss of chemical is determined and then tested at least monthly. Samples for this purpose should be taken from a suitable connection on the steam drum, if possible.

(1) Flush the line thoroughly to assure that the sample represents internal boiler water. Avoid excessive removal of water from the boiler.

(2) For complex systems, if testing shows that the volatile oxygen scavenger concentration decreases 33% from the original concentration or the pH of the boiler water drops below 9.0, additional chemicals must be added to the boiler to re-establish the desired concentrations.

(3) Water should be recirculated long enough to be sure that all chemicals are well mixed and an analysis made for the inhibiting chemicals. This should be done on a weekly basis. The recirculation should last long enough to move three to six volumes of the lay-up solution or for a minimum of 10 minutes.

(4) A practical procedure may be to add additional lay-up solution, circulate by lowering the water level and firing lightly, and then follow the procedures previously described, as applicable, for complete filling of boilers.

Attention should be directed to valve maintenance that may cause treated water loss or the addition of untreated water into the boiler and auxiliaries causing dilution of properly treated lay-up solutions.

### ❏ 14.2 When An Inert Gas Is Used

All pressure gauges should be monitored on a weekly basis to verify that leaks have not developed. The inert gas pressure should not be allowed to drop below 5 psig (34 kPag).

### ❏ 14.3 Equipment Sealed And Maintained Dry With Desiccants

Equipment with desiccants in visible reservoirs should be checked once a month. Silica gel that has changed color or lime that has formed lumps should be replaced. In the case of sealed equipment containing desiccant, if the seal is broken, the equipment should be opened and the desiccant replaced. If signs of moisture are found when the equipment is opened, the equipment should be dried out before adding fresh desiccant.

# ❏ Section 15 ❏
# START-UP AFTER LAY-UP

It is essential that proper procedures be followed for restarting equipment after lay-up. The effectiveness of the lay-up may become apparent during start-up, if the start-up is adequately monitored. Indicators may be very high or very low crud concentrations entering the boiler from the condensate / feedwater system. If satisfactory care was not taken, equipment recleaning or even replacement may be necessary before returning the equipment to service.

Preparation for start-up should begin well in advance of the projected start-up date. Some suggestions to help simplify start-up are:

(1) Before start-up, determine the parameters which must be monitored and ensure that personnel and/or equipment will be available to monitor these conditions adequately.

(2) Any steam or water, that is not part of the lay-up procedure should not enter the equipment prior to start-up. If steam or water does enter, it will likely remain and combine with gases or other impurities to form a corrosive mixture.

(3) All blind flanges and sealing compound must be removed from the turbine and auxiliary systems.

(4) The equipment should be inspected to determine if post lay-up cleaning is necessary. Valves, motors, pumps, and turbine blading should be inspected. If corrosion products or deposits are found, they should be removed. The circulating turbine oil should be cleaned and checked for corrosion inhibitor, microorganisms, biocide level and particle count.

(5) If start-up is delayed after inspection, conditioned air should be circulated through the system using the same procedure as described in "Recirculating Dry Air" (p.31).

(6) The manufacturer's start-up procedures should be followed.

(7) The impact of lay-up chemicals on start-up techniques and subsequent operation must be considered from an environmental, health, and equipment preservation point of view. If the chemicals used during lay-up are to be drained and disposed of prior to start-up, the chemical supplier should provide an assessment of the environmental and personnel toxicological concerns during disposal.

(8) On-site wastewater personnel and/or local regulatory agencies should be notified prior to disposal of lay-up chemicals.

(9) If lay-up chemicals are to be left in the system during start-up, the chemical supplier and equipment manufacturer should be consulted to determine the acceptability of the chemical and the potential for undesirable side effects such as polymerization or degradation to corrosive intermediates. If the potential for steam cycle equipment corrosion during start-up exists, a program should be implemented to counteract or minimize the anticipated effects.

(10) It is important that all desiccant materials used be removed prior to start-up. If the desiccant material has been spilled, the boiler must be thoroughly cleaned by vacuum or other suitable means and flushed with condensate, or better, quality water before start-up.

# ❐ Section 16 ❐
# CONCLUSIONS

Each outage, depending on its length and nature, will present a unique set of problems. Therefore, all lay-up programs should be designed to overcome such problems and yet remain flexible enough to accommodate unforeseen events and situations that may arise. The basic lay-up procedures in this consensus are intended to present the methods and materials necessary to meet this need.

# GLOSSARY OF TERMS

*Breach of system* —system is opened and air ingress occurs on the waterside (see definition below).

*Complex circuitry* — steam circuits which include one or more non-drainable components including superheaters, reheaters, preheaters and economizers.

*Condensate-quality* — condensed steam with specific conductance (excluding the contribution from ammonia or amines) equal to or better than that of the demineralized water specification below.

*Deaerated water* — containing $\leq$ 7 µg/L (ppb) oxygen as $O_2$ measured before chemical oxygen scavenger addition.

*Demineralized water* — treated water produced by resin or membrane processes having a specific conductance $\leq$ 10 µS/cm (µmhos) at 25°C and containing $\leq$ 0.01 mg/L (ppm) iron as Fe. For use in non-drainable superheaters or reheaters constructed with austenitic stainless steels, this specification shall be $\leq$ 1.0 µS/cm (µmhos) at 25°C.

*Dehumidified air* — air with less than 30% relative humidity at ambient temperature.

*Dew point* — the highest ambient temperature at which condensation will occur.

*Filtered water* — treated water containing turbidity $\leq$ 1 NTU, suspended solids $\leq$ 1 mg/L (ppm) and color $\leq$ 5.0 APHA.

*Long-term lay-up* — varies with equipment type. See text in applicable equipment section.

*Mothballing* — equipment is permanently removed from service but maintained in a preserved state for possible return to service if ever required

*No breach of system* — no air ingress occurs on the waterside.

*Preoperational post hydro* — the period following hydrostatic testing and preceding chemical cleaning of utility boilers

*Quicklime* — 90% Calcium Oxide (CaO)

*Short-term lay-up* — varies with equipment type. See text in applicable equipment section.

*Simple circuitry* — all steam cycle components, including superheaters, reheaters, preheaters and economizers, are fully drainable.

*Waterside* — all steam cycle components, including superheaters, reheaters, preheaters and economizers, that come in contact with water and/or steam.

# ◻ REFERENCES ◻

1. ASME Research and Technology Committee on Water and Steam in Thermal Systems, Boiler Lay-up Task Group, "Consensus of Current Practices for Lay-up of Industrial and Utility Boilers," ASME HO 00336, 1985.
2. Beecher, J., and Lane, R. W., "Guidelines for Lay-up of Boilers and Their Auxiliaries," *Proceedings of the International Water Conference,* Vol. 42 , 1981, pp. 155–158.
3. *Steam / Its Generation and Use*, 38th edition, Babcock & Wilcox Company, 1972.
4. *Combustion Engineering, a Reference Book on Fuel Burning and Steam Generation*, revised edition, 2nd Impression, Combustion Engineering, Inc., 1967.
5. Armentano, J. A., and Murphy, V. P., "Stand-By Protection of High Pressure Boilers," *Proceedings of the International Water Conference,* Vol. 25 , 1964, pp. 111–124.
6. American Society of Mechanical Engineers, *ASME Boiler and Pressure Vessel Code, Section VII, Recommended Rules for Care of Power Boilers,* Part C7.300, Laying-up of Boilers, 1983 edition.
7. Reid, W. T., "Protecting Standby Equipment, Corrosion Control During Equipment Shutdown," *Materials Protection*, Vol. 6, July 1967, pp. 42–44.
8. American Society of Mechanical Engineers, *The ASME Handbook on Water Technology for Thermal Systems*, Chapters 21 & 22, pp. 1705–1743, ASME I00284, 1989.
9. Pike, T. H., "Corrosion Prevention of Turbines During Extended Outages (Case Histories)," *Proceedings of the International Water Conference,* Vol. 48, 1987, pp. 200–208.
10. Pike, T. H., "Corrosion Control of Condensers, Feedwater Heaters, and Turbines During Idle Periods," *Corrosion/89*, National Association of Corrosion Engineers, Paper 73, New Orleans, LA, April, 1989.
11. Daniels, D. G., "Follow Good Lay-up Practice to Prevent Cycle Corrosion," *Power*, Vol. 142, No.2, March/April 1998, pp. 37–44.

12. Hopkins, R. D., Hull, E. H., and Shields, K. J., "Guideline Manual on Instrumentation and Control for Fossil Plant Cycle Chemistry," Electric Power Research Institute Report CS – 5164, Sections 5 & 6, April 1987, pp. 5 - 1 to 6 - 14.
13. Daniels, D. G., "Good Lay-up Practice Produces Significant Benefits," *Proceedings of the International Water Conference*, Vol. 59, 1998, pp. 367–373.